DECO DESIGN
Miami Beach Style

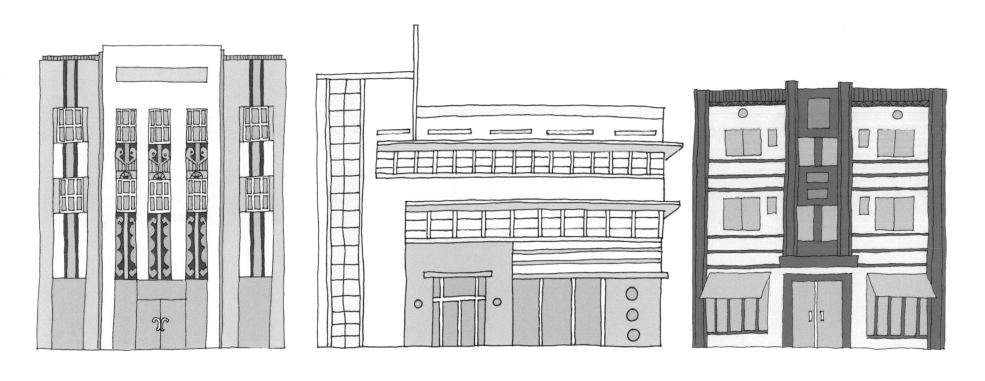

Isabel Hill

To my dear friend Jan,
a fellow traveler in my Miami Beach adventures!

Please visit www.isabelhill.com

ISBN: 979-8-9858567-2-9

Book design and illustrations by Catherine Hnatov

Library of Congress Control Number: 2022904828

It was the 1920s when Art Deco came about.

Its simple characteristics made it easy to pick out.

In Miami Beach, buildings have a special flair.

Colors, lines, and shapes are the elements they share.

In this place on the seaside, blue and sandy,

buildings are the **color** of cotton candy.

To shade the bright sunlight in this tropical place,

 **eyebrows** above windows add an element of grace.

On the top of this building, there's a Deco motif.

Reminding us of the beach is a **wave relief**.

They look like portholes on a very large boat.

 Octagonal windows are a detail to note.

On a corner medallion, carefully curved;

 a geometric fountain is beautifully preserved.

Like waves coming ashore, they do not connect.

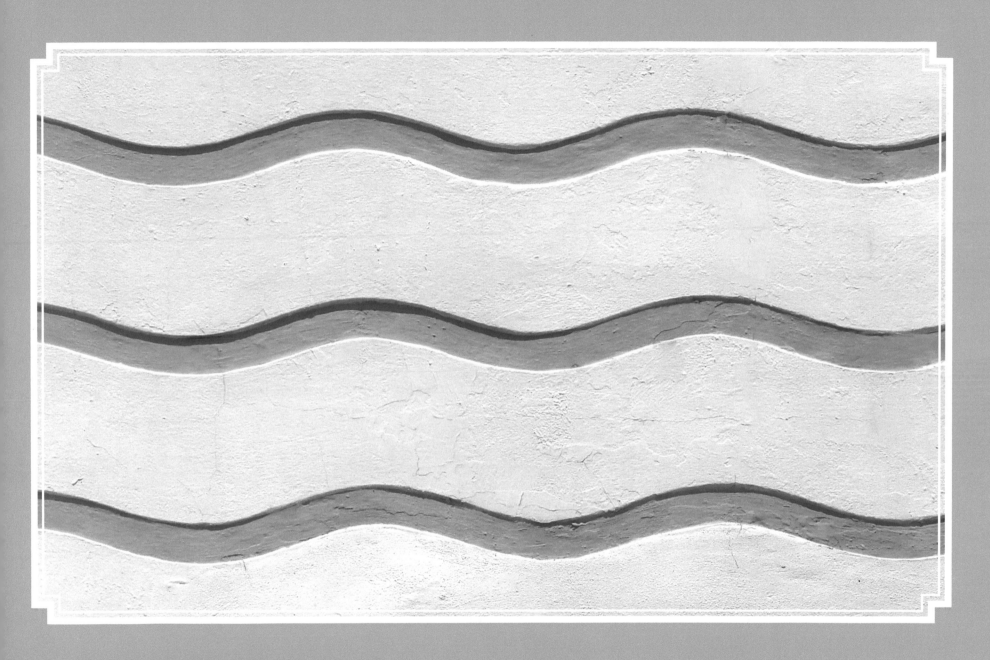

 Horizontal lines make this Deco effect.

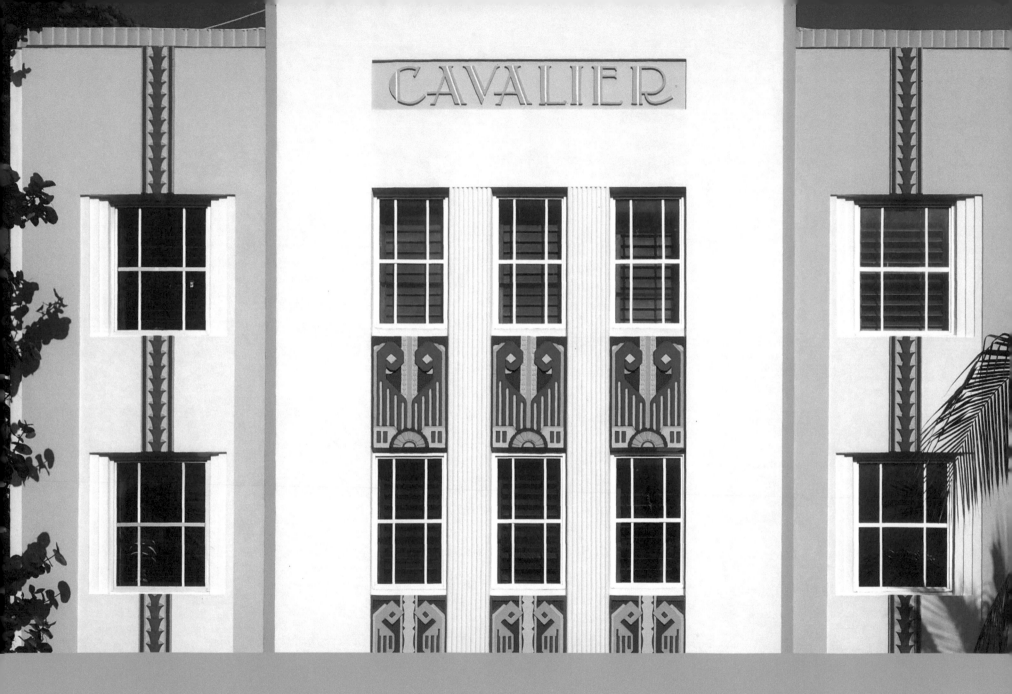

Spirals, lines, and circles on a matching tile,

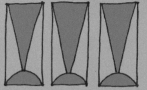 details come in **threes** in the Art Deco style.

 Long, shiny, and **streamlined**, it looks like a train.

Original details of this diner remain.

Thick purple edges rising to the roof line

meet a horizontal band with **zigzag** design.

Art Deco design comes in different shapes and types.

This building is square with vertical stripes.

A long, curved building looks like a **ship**,

with shiny metal railings, another Deco tip.

The materials that are used can also give a clue.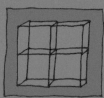

Glass block is a Deco detail that you cannot see through.

Horizontal windows make the building look wide.

When **vertically** stacked, they stretch up one side.

This rectangular building has a roof that is flat,

and stepped up sides make a **ziggurat**.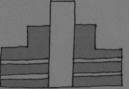

Circles and squares are shapes to look for.

At the entrance, you might find a **terrazzo** floor.

The name of each building is shown on a sign.

Simple bold lettering creates a strong line.

Blue, pink, orange, green, silver, or white—

colors change with **neon** when day turns to night.

Go back to each building and try to combine

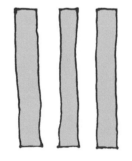

colors

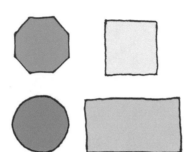

octagon, square,
circle, rectangle

eyebrows

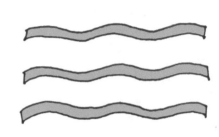

horizontal lines

vertical lines

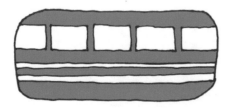

streamlined shape

zigzag design

wave motif

all the Art Deco details that you can find.

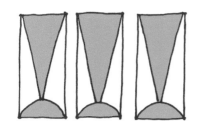

designs in threes

ship-like design

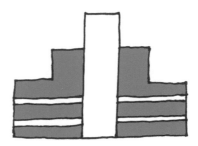

ziggurat

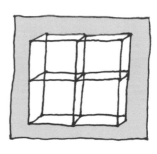

glass block

terrazzo floor

simple bold lettering

fountain motif

neon

Deco Destinations

Webster Hotel
1220 Collins Avenue
Built: 1936
Architect: Henry Hohauser

Carlyle Hotel
1250 Ocean Drive
Built: 1941
Architects: Kiehnel and Elliott

Marlin Hotel
1200 Collins Avenue
Built: 1939
Architect: L. Murray Dixon

Park Central Hotel
630 Ocean Avenue
Built: 1937
Architect: Henry Hohauser

Bentley Hotel
510 Ocean Drive
Built: 1939
Architect: John and Carlton Skinner

Whitelaw Hotel
808 Collins Avenue
Built: 1936
Architect: Albert Anis

Cavalier Hotel
1320 Ocean Drive
Built: 1936
Architect: Roy I. France

Eleventh Street Diner
1065 Washington Avenue
Built: 1948
Builder: Paramount Dining Car Company
of Haledon, New Jersey

Shelley Hotel
844 Collins Avenue
Built: 1931
Architect: Henry J. Moloney

Leslie Hotel
1244 Ocean Drive
Built: 1937
Architect: Albert Anis

Sherbrooke Apartments
901 Collins Avenue
Built: 1947
Architects: MacKay & Gibbs

Waldorf Towers Hotel
860 Ocean Avenue
Built: 1937
Architect: Albert Anis

Kent Hotel
1131 Collins Avenue
Built: 1939
Architect: L. Murray Dixon

Breakwater Hotel
940 Ocean Drive
Built: 1939
Architect: Anton Skislewicz

Essex House
1001 Collins Avenue
Built: 1937
Architect: Henry Hohauser

Crescent Hotel
1420 Ocean Drive
Built: 1938
Architect: Henry Hohauser

CPSIA information can be obtained
at www.ICGtesting.com
Printed in the USA
BVHW092105120123
656185BV00006B/47